Material Culture and Technology

ROBERT F. G. SPIER

University of Missouri
Columbia, Missouri

Burgess Publishing Company • Minneapolis, Minnesota

Printed in the United States of America
Library of Congress Card Number 73-82851
SBN 8087-1970X

1 2 3 4 5 6 7 8 9 0

A SERIES ON
BASIC CONCEPTS IN ANTHROPOLOGY
Under the Editorship of
A. J. Kelso, University of Colorado
Aram Yengoyan, University of Michigan

Contents

Some Definitions

Material culture forms the most obvious part of culture. Simple as a digging stick or complex as a moon rocket, each item of material culture is the end product of culturally directed behavior. Each is reflective of the culture which gave rise to it.

For the sake of convenience in thought, anthropologists customarily distinguish those manifestations which have a physical presence from the total culture. The former have been called "material culture," while the remnant portions are usually not designated other than by the term "culture," which is also applied to the whole. The interest in material culture arises in part from the possibility of collecting and arranging these objects, something which is difficult with nonmaterial culture. Consequently, material culture has usually been the business of museums and their inhabitants.

It is most important that we do not give artifacts a life of their own. They are the results of behavior and are created in no other way. In some respects one may maintain that there is no difference, culturally speaking, between the forms of baskets, for example, and the forms of postmarital residence. We tend to view them differently for reasons which are irrelevant, such as the physical nature of basketry versus the abstract nature of residence patterns. Material culture, then, is the concrete result of human learned behavior.

Technology, on the other hand, is the complex of learned behaviors which gives rise to material culture. Again we are dealing with an aspect of culture. The knowledge, attitudes, and customs of technologies are as much a part of the cultural baggage of man as his political or religious behavior.

Technology embraces the means which man uses to modify his natural environment. However, in discussing material culture, we exclude from our definition of technology some means of environmental modification. One of the exclusions is the endeavor to enlist the assistance of supernatural forces in altering the state of nature. The efficacy of prayer, sacrifice, or similar processes toward these ends may be questioned. Yet it would be false to insist that the means be rational (whose rationality?), or physical (virtually all approaches to the supernatural involve material objects—amulets, sacrifices, etc.), or even effective (have modern chemical preparations rid your lawn of weeds?). We will also exclude at present environmental modification through social means. A definite fact of one's environment is population density. Density may be limited in modern, urban societies by property zoning. It may also be increased in the high-rise housing project, which is both a physical artifact and a social event. Consideration of these two means of modifying the environment would substantially complicate our task without, it would seem, offering compensatory understanding.

For our purposes, then, the result of technological activity is the creation of artifacts, the stuff of material culture. While the nature of material culture is readily understood—for everyone recognizes man-made things—the several usages of the term "technology" often lead to confusion. "Technology" applies to human activities on several levels of abstraction. We may use it, most grandly, with reference to the organization and content of all man's technical pursuits at all times. Or from this manwide scope we may retreat to one time and one place (a more usual reference) as in "Western European technology." Or more specifically, we may speak of individual or subsidiary technologies, such as the "subsistence technology" or "ceramic technology." Obviously, one becomes more concrete and particular in reference toward the lower end of this scale.

Each actual technology, however, regardless of the level of abstraction in which it is treated, is composed of a series of coordinated techniques. In a given circumstance these techniques are directed toward a particular goal as the situation is evaluated by the members of that culture. The chosen techniques are not necessarily those which might be employed by members of another culture seeking the same goal. For example, the surface deterioration of a wooden artifact might be handled, in one culture, by scraping or sanding to develop a fresh surface; in another culture painting over the blemished surface might be the treatment.

Though this preamble is couched in terms which suggest ancient and preindustrial cultures, it must be realized that the modern world grew out of the past. The content of material culture and technology show surprising continuity, and the dynamics are much the same in both ancient and modern cultures. For virtually every situation discussed, modern as well as ancient or primitive examples may be found. The same principles apply.

Pathways to the Present

The study of man's technology reaches to times which are older than man himself, paradoxical though this may seem. At one time anthropologists were quite rigid about tool making and using, holding them to be exclusively human behaviors. Thus man was defined in terms of tools: tool users were man; only man makes and uses tools (Dart 1959: 217; Oakley 1972: 1-3). However, it has become increasingly apparent that tools were made and used by nonhuman creatures, some of which, like Darwin's finch, are only remotely related to man. Studies of nonhuman members of the Primate Order have shown substantial tool use, also (Pfeiffer 1969: 48-51). Therefore, one may not use tool use or tool making as touchstones to determine humanness. The difference between man and animal in this regard must lie in degree rather than kind, for at least the humans would insist on the existence of a difference. The response is that man, no matter how humble his life, uses more tools more often than do animals. Man, possessed of this extrabiological means of rapid adaptation, has now become dependent upon tools, not any given tool but the fact of tools generically.

Under the tool-using condition, a new assortment of human physical traits was permitted to persist. It is likely that some of them, such as erect walking, were positively adaptive for tool use; erect posture freed the forelimbs for carrying and manipulation. Other physical traits, such as the reduction in jaw size, were neutral with respect to tool use but were likely maladaptive in the absence of tools. Given the use of tools the size of the jaw was of no consequence, and a reduction is observed to have occurred. Without tools man's jaws and teeth would be employed as weapons (even just to display as a threat, as many other Primates do) and probably as tools to sever, sharpen, split, and otherwise modify materials other than food. In general, the newly established artificial environment (which we call "culture") was a context within which greater variation in man could be tolerated. At another time, in the absence of culture, some of these same variations would have been disadvantageous and their bearers probably selected out.

The impression should not be left that all of early culture consisted of tool making and tool using. We must assume that other learned, shared behavioral traits existed. However, the archeological record on which we must depend for evidence offers us little beyond the remains of the tools themselves, the products of behavior. Insofar as these tools conform to a definable series of styles—possess a finite number of features in well-established combinations—we can consider them the results of cultural behavior.

Although some doubt has been expressed about the human origin of some bone items which are allegedly tools of Australopithecines, most stone tools have a ready acceptance as such. Once one passes beyond the earliest times of man's existence, there is little doubt in the minds of most observers that technological products are the creation of man. Wide experience with the cultural products of other times and places leads to increasing confidence of identification. There are two interesting sidelights to this point. First, if you contemplate the matter for a moment you will realize that you can identify as man-made an object of whose use you are ignorant. For example, the purpose of a stone object, resembling a large wing nut, called a "bannerstone" was unknown to archeologists, but the item was unquestionably cultural (Martin, Quimby, and Collier 1947: 37-38, fig. 69). Another case of dubious identity was that of the tread-trap. This wooden artifact, consisting of a solid body with a rectangular "window" crossed by loose rods, was variously thought to be a trap, a musical instrument, or models of boats. A fortuitous archeological discovery plus ethnographic data has revealed its true nature (Clark 1964: 187-188). Second, sometimes the scale of the product obscures its human origin. For years some of the great Indian mounds of the Mississippi River Valley were considered to be natural hills. The artifacts found on them were deemed only evidences of Indian occupation. Recent archeological borings and trenching have demonstrated that these features are wholly artificial, built up from level ground by hundreds of thousands of basketloads of dirt (Jennings 1968: 216-222). One marvels at the persistence evidenced.

The development of technology is customarily traced through a series of traditional periods or Ages. (The scheme is modelled after the developmental sequences used by geologists and paleontologists.) When the system was first enunciated in the early nineteenth century, it fitted well with the unilinear cultural evolutionary theories of the times. With each technological stage there was thought to exist, inevitably, a given stage of social organization, of religion, of family organization, of political sophistication, and of other aspects of culture. Today we do not follow such a developmental scheme as narrowly as before, but there has been a broad resurgence of evolutionary thinking about culture, associated in part with the work of Leslie A. White and his students (White 1949; Sahlins and Service 1960). In its modified form the concept of general evolution has become widely acceptable. There is agreement on the main trends of human culture, on the associations between various parts of culture, and on the necessity of preconditions to underlie subsequent cultural steps. The following comments are offered in the light of this newer view of cultural evolution with a realization that the schema will not follow in every instance which may be cited. The view serves only as a general guide to thought.

The customary named Ages of prehistory are Paleolithic, Mesolithic, Neolithic, Copper, Bronze, and Iron. The first three are sometimes called the Old, Middle, and New Stone Ages and the latter three the Metal Ages. As will be seen below, the difference that exists between the first two Ages and the latter four is probably the most significant. The Ages are often divided internally into Lower, Middle, and Upper parts (respectively, early, middle, and late).

THE PALEOLITHIC AGE

The Paleolithic Age begins with the earliest evidences of man's culture which we discussed above, largely stone tools of initially crude types. Percussion techniques were the first used to chip stone into desired shapes. Either waste flakes were removed from a matrix ("core") to leave the tool as the remainder, the "core tradition" of stone working. Or large flakes were removed from the core, and these flakes themselves were worked into tools, with the remaining core as the waste, the "flake tradition." The percussion blows were struck directly with a hammer stone, which being handleless was held in the hand. For heavy chipping the workpiece might be struck against a stone "anvil" to remove large flakes. The blow of stone upon stone is sharp and shattering so as to produce only crude results. By the Middle Paleolithic the percussion techniques were augmented by the use of bone and antler punches, to cushion and direct the blow of the stone hammer, and by hammers of these same nonlithic materials. Greater control was gained and finer chipping resulted. Work continued in both core and flake traditions.

The Middle Paleolithic, obviously a time of innovation in stone working, saw the advent of a second major approach, that of pressure flaking. In this technique a pointed piece of bone or antler (a "flaking tool") is pressed forcefully against the stone in such a manner that a flake is driven off.

Direction, force, and abruptness of application may be varied to give additional control over results. Pressure flaking, of primary application in the flake tradition, is usually used after initial percussion chipping of the workpiece. Its use led eventually to some work of a high order of artistry in the European Upper Paleolithic and elsewhere around the world in more recent times. These stone tools may be admired as being the masterpieces which they are.

Another Middle Paleolithic innovation was the development of the prepared core approach (basically in the flake tradition) as reflected in, for example, the Levalloisian technique. The stone tool was percussion-chipped into substantially its final shape while still attached to the parent stone ("tortoise" core) on which it was formed. A final, parting blow detached the completed tool from its bed and defined its inward face at the same time. The prepared core approach made possible the fine blade tools of the Upper Paleolithic and later times.

High-quality stone chipping depends greatly upon high-quality raw materials. Stones possessing a "conchoidal" fracture ("shell-like," with a cupped curvature like a clam or oyster shell) such as obsidian, flint, chert, chalcedony, and jasper lend themselves to percussion and pressure techniques. Stones with marked lines of cleavage, like slate, or those with an even, granular structure, like quartz or sandstone, are only with difficulty chipped into tools. These latter, smooth-grained stones can be shaped by pecking and grinding techniques such as those which came into use in the Neolithic Age.

Much of our knowledge of Paleolithic stone working is derived from a combination of observation and experiment. New investigative devices, such as the electron microscope and the photomultiplier, have made possible refined observations on the artifacts from which we have learned details of both manufacture and use (Semenov 1964; Purdy and Brooks 1971). For decades amateur archeologists, and an occasional professional, have been trying their hands at duplicating the stone artifacts of primitive and prehistoric man. This imitation has lately turned to controlled experimental study which is reconstructing, with some certainty, the techniques of native stone workers (Bordaz 1970; Bordes 1968: 15-31; Leakey 1954; Crabtree 1970). As with the scientists of any age, we are confident that we have the right answers to old questions. We might even be correct.

Aside from the stone, other cultural evidences from the Lower and Middle Paleolithic are scant. Apparently many peoples of this Age occupied campsites with temporary shelters whose contents have often been scattered and destroyed. Cave dwelling was on the rise in the Middle Paleolithic, leaving, in this protected context, a better record of the domestic life and industrial habits of early man. Though organic materials have generally not survived, the stone tools associated with their conversion into artifacts afford inferences about their presence. We are led to these conclusions by the shapes of cutting and scraping tools, together with the edge wear resulting from use (Semenov 1964; Keller 1966).

The Upper Paleolithic saw the flowering of a life style which had been several million years in the making. The crude hand axes of earlier times virtually disappeared in favor of finely chipped products which often owed a developmental debt to the prepared core techniques. The tool inventory was much enlarged, with numbers of specialized tools present. For example, scrapers of every description—side scrapers, keeled scrapers, end scrapers, double-end scrapers, and so on—abounded and testify to extensive working of wood and hides. Gravers (also called burins), found in numerous forms, were used in making antler and bone artifacts, such as pins, needles, spear points, fish gorges, and harpoon points (Oakley 1972: 59).

Lastly, but most importantly, the long Paleolithic was a time during which man did not cause or stimulate the growth of food; he relied solely upon that which was available in nature. This was the era of food gathering or food collecting. The processes involved might be hunting, or fishing, or fowling, or trapping, or gathering, or some similar technique. The richness of the environment, the efficiency of exploitation (which hinged in part on tools), and man's diligence all governed the resulting subsistence. When all factors were favorable the exploiters might live very well indeed. They were then able to act in many respects like peoples of the following cultural Ages when man produced his food.

Two classic illustrations of great success in food collecting are the cases of the Upper Paleolithic (Magdalenian) inhabitants of southwestern France and the nineteenth-century Indians (Tlingit, Tsimshian, Haida, and Kwakiutl) of the Canadian northwest coast. The Magdalenians lived seasonally in caves and in open sites. They hunted the plentiful cold temperate climate animals of their region—horses, bison, wild cattle, reindeer, ibex, and others. They had sufficient time, beyond the demands of everyday living, to produce cave paintings, engravings, and low reliefs which rank, even today, as high artistic endeavors (Powell 1966: 9-63; Laming 1959; Kühn 1955). The Northwest Coast Indians lived in permanent villages of large wooden houses facing seaward along a beach. They fished, both offshore and on the nearby rivers, hunted land animals and waterfowl, and gathered roots and berries in season. The abundance and reliability of seafood made possible the accumulation of surpluses. Wealth was correlated with marked status differences, reaching from nobles on one hand to slaves on the other. High-ranking individuals sought to enhance their social position through a system of competitive feasts in which host sought to humble guest by lavish display of generosity and conspicuous consumption of goods. These "potlatches," for which these Indians have become justly famous, used great quantities of food, both eaten and wantonly destroyed, and goods, such as canoes and blankets, which were the products of food (the producer ate while he produced) (Drucker 1965: 55-65). All of this ostentation and consumption was sustained by a mode of subsistence in which theorists place little faith, but which is upon occasion capable of great results, as we have seen.

More commonly the food gatherers got by reasonably well, with seasonal gluts and famines well known to some. At the least pleasant end of the scale there were people who doubtless worked long, hard, and efficiently to wrest a bare living from a reluctant land. Some modern primitive peoples, mentioned below, are good candidates for this last category.

THE MESOLITHIC AGE

The Mesolithic Age which followed has been termed a post-Pleistocene continuation of the Paleolithic (Braidwood 1967: 81-87; Binford 1971: 27). About 10,000 years ago the last Pleistocene glacial advance began its retreat and the animals and plants adapted to a cold climate gave way in many places to those which thrived in temperate and even subtropical climates. The men of this time were stimulated to alter their cultures accordingly, but many of them clung to some form of food gathering. The primitive hunting peoples of the modern world, such as the Athapaskans of northern Canada, the Australian aborigines, and the Bushmen of southern Africa, are typologically Mesolithic peoples.

Contrary to the examples given above, the demands of food gathering usually place limits on the elaboration of material culture. While the food quest is not an unremitting round of toil, it does take a great deal of time, with seasonal peaks and lows. This life frequently involves seasonal movements and temporary residence away from the usual domicile. For this travel the food-gathering people, with a few notable exceptions (such as the reindeer-driving Siberian hunters), have as transport facilities only the human head, hand, or back. The consequences are that these people have limited time for craftwork, can't readily engage in lengthy processes or those which require cumbersome equipment, and are restricted in the bulk, weight, and quantity of goods they can possess and move.

The end of the Mesolithic did not come uniformly with the first appearance of the Neolithic Age. In some instances the Mesolithic peoples may have been well satisfied with their way of life and loath to exchange it for another. In other cases the mode of livelihood initially implicit in the Neolithic subsistence type was ill-adapted to the environment and required substantial refinement before it became attractive. And in all cases the product of change from Mesolithic to Neolithic was far from uniform. The seed grew, almost literally, in a way dependent upon the cultural soil in which it was planted.

THE NEOLITHIC AGE

The Neolithic Age saw the advent of a revolutionary subsistence economy, namely, food production. Men began to create new food supplies, to augment and assist natural food supplies, and to modify these foods to suit human convenience. The Neolithic Revolution bears this name because of the potentiality for great change, not because of its speed. Although plant cultivation may have started as early as 8000 B.C. in parts of the Near East (and not much later, and independently, in Southeast Asia and Middle America), it took thousands of years for its major impact to be apparent. The assumption of control over the food supply had tremendous consequences for population location and density, for the creation of surpluses and accumulation of wealth, and for occupational specialization, to enumerate but a few potentialities.

Food gatherers had always been limited in their food to that which grew or lived in a given zone of exploitation. To be sure, hunting lands might be expanded to meet the needs of a growing population, but there was a definite limit to this process as territories impinged upon each other and as the hunter got farther and farther from home base. Further, as the population grew, the presence of humans in large numbers tended to drive away some game. More intensive exploitation of a fixed area would give higher yields, but could not be continued season after season without destroying the plant and animal populations by exceeding their reproduction rate. By contrast, the food producers could raise more food on an area than grew there naturally and consequently could support more people on less land. Their activities, too, had limits, such as arable land available and distance from home to garden, but these were much higher than the limits on the food gatherers. Only in the last few centuries with a sharp rise in world populations has there been widespread evidence of land shortage on a global scale.

Not only could food production be expanded to meet the needs of expanding populations, it was capable of outstripping these needs to produce surpluses. Plants could be produced which were not essential foods, but were instead the basis for stores of wealth, for feasting, and for the fibers, oils, and other materials of craft activities. In a less marked way, plentiful supplies of animals contributed to this situation. It can be understood that the ability or opportunity to command wealth leads to power and to social distinctions. The ability to stage feasts will gain the goodwill or allegiance of others. And the potentiality of raw materials in large quantities for crafts (or, later, industries) is related to occupational specialization as those materials were utilized.

Aside from its subsistence technologies, the Neolithic showed a change in techniques of stone working. Characteristic tools of this Age were produced, or finished, by grinding and polishing rather than by chipping. Initial shaping often involved chipping or pecking (crumbling away small particles), but the final finish to the cutting edge or overall was by grinding. In the instance of stone axes the ground edge was more effective in wood cutting than had been the chipped edge of the hand axe. This advantage was important in clearing fields for cultivation.

The differences between stone tools made by chipping and those made by grinding have been suggested as reflective of differences in the psychology of the stone worker (Kroeber 1948: 629-630). In the first place we note that good stone chipping calls for greater skill than does stone grinding. Further, stone chipping is much quicker than grinding; it is possible to make a serviceable knife, graver, scraper, axe, or projectile point in less than a half hour. Grinding takes hours, days, or even weeks for comparable products. Kroeber proposes that the late appearance of the evidently simpler technique is related to the ultimate development of self-discipline, the deferment of gratification. Lastly, on a practical level, the production of any tool which involves much labor is doubtless related to an assured food supply while at work and the prospect of keeping the result by reason of fixed residence or ready means of transport.

The Neolithic communities were larger than those of previous Ages, the comparison being between groups numbering in the tens or a few hundreds and those counted in scores or many hundreds. Settlements were often occupied throughout the year and used for decades or generations. Houses and auxiliary structures were built for the long haul. Stone vessels—mortars, bowls, lamps—increased in number. Pottery became a characteristic product of these sedentary peoples. (So much was this the case that some prehistorians consider all pottery makers to have been food producers and vice versa.) Weaving of textiles became common, with cultivated plants providing the requisite vegetable fibers and domestic animals the wool and hair.

Anthropologists often make a formal distinction between two food-producing modes: horticulture and agriculture. Horticulture is plant cultivation making use of hand tools such as the hoe (sometimes resulting in the name of "hoe culture"), the spade, or the digging stick (Curwen and Hatt 1953: 62-64). Agriculture is plant cultivation with the plow, implying the use of animal traction to draw the implement. A third food-producing mode, less widely found than either horticulture or agriculture, is pastoralism which relies on the exploitation of domestic animal herds. In practice, horticulture and agriculture rarely occur as pure types but occur as mixed farming, with domestic animals involved as well.

Horticulture, even though given one name, is not everywhere the same set of techniques. In the first instance, each culture tends to have its own collection of plants and to favor the use of one tillage tool or a definite group of tools (Kramer 1966). As you can well imagine, there are differences between cultivation with the spade and that with the hoe. Dry rice, potatoes, and maize (Indian corn) were all grown by methods termed horticultural, but the techniques for one crop could hardly suit them all.

Many horticulturalists cultivated lands, such as those in river bottoms, which later were directly convertible to agriculture. However, many other horticulturalists worked on lands which were completely unsuited to the plow because of the difficulty of cleaning. In forested lands slash-and-burn cultivation (also known as swidden or milpa) used temporary clearings for a few years' crops before encroachment of grasses and other second growth forced abandonment. The horticulturalists planted their crops in small patches and soil pockets between stumps and fallen trees. The clearing process was repeated in the vicinity while the original plot was abandoned for years until the forest grew back. It has been estimated that such methods required access to seven to ten times as much land as was cultivated at a given time (Curwen and Hatt 1953: 214, 285). With good fortune a village might stay in one spot and continually rework the surrounding land. Often a village moved, perhaps once a generation, to a new forested area when its lands did not regenerate quickly enough to keep the fields in convenient reach (giving rise to the terms "shifting" or "migratory horticulture"). As long as overall populations were small and the lands large this technique presented few problems. The technique is still in vogue in parts of New Guinea and the Amazon Basin.

The generalized picture of the Neolithic which we usually draw, one of village farming, is based upon the Near Eastern grain farmers who had some animals at an early time and quickly, technically, became agriculturalists. Thanks to irrigation, manuring of fields, crop rotation, fallowing, and the renewal of soils by periodic flooding these people were able to cultivate in substantially the same places for millennia. However, what is true of them does not apply with equal force everywhere. A thorough discussion of the modes of life of horticulturalists, agriculturalists, and pastoralists is to be found in Forde's *Habitat, Economy and Society* (1963).

In conclusion, the Neolithic Age saw a substantial forward movement in the production of material culture. Sedentary residence, combined with more time away from the subsistence quest or more food surpluses to pay for someone else's time, resulted in more goods per capita. There remained some poverty-stricken people who barely wrested a living from their habitat, and there were others who were sucked dry by the demands of rulers and landlords, but generally more people had more things around them.

THE COPPER AGE

The Metal Ages were not, for a long time, as revolutionary in their nature as had been the Neolithic. While they eventually opened new avenues for cultural development, the initial result was to intensify trends already identifiable in the fully developed Neolithic. The growth of population and its density, the furtherance of occupational specialization, the increasing disparity between the rich and the poor, and the dominance of technologically elaborated peoples over their neighbors with less complex material culture are all observable tendencies.

The Copper Age (also called Chalcolithic) is almost as early as the Neolithic if we consider as its origin the most ancient evidences of man's metallurgical activities. Copper, like gold and silver, is sometimes found in a natural metallic state. Men's first use of the metal, early in the sixth millennium B.C., was in this native form rather than smelted from an ore (Wertime 1964: 1258). Consequently, the earliest metallurgical treatment of copper was not smelting but annealing, a process employed to reduce work-hardening and permit further shaping of a workpiece. In unalloyed form this metal may be hardened to a degree by hammering but generally must be classified as a soft metal. For this reason, as soon as harder metals were available, its principal use was ornamental—in rings, pins, bracelets, and other pieces of jewelry.

A Copper Age of utilitarian type didn't really come into existence until a full-fledged Neolithic had held sway for a while. Despite its shortcomings, copper did come to have its uses in weaponry and tools. Socketed axes of complex form were made by casting in pieced molds (Danubian axes are illustrated in Childe 1925: 188). Hammered copper daggers had a broadly triangular outline, in fact so broad as to approach an equilateral triangle, suggesting that the metal was not strong enough to fashion a narrower blade. Rising nations accumulated hoards of old copper articles from which weapons were made as needed. Possession of these slightly superior armaments evidently gave them the edge, so to speak, over their lithic neighbors. The habit of the time to hoard and rework metals has had its archeological consequences. These hoards are occasionally found and prove to be great treasure troves of junk (literally), but must be considered the counterparts of our stockpiles of strategic materials. In the absence of such finds the earlier artifacts would have disappeared from the scene, much as older model cars are scrapped today to make new ones.

The Copper Age was not a revolutionary period and did not produce spectacular results. It was a time of learning about metallurgy, a time of perfection of food-producing techniques, and a time of the inception of substantial political units. Some prehistorians prefer to pass over the Copper Age, along with the Mesolithic Age, because of its transitional nature. This appraisal may be correct.

THE BRONZE AGE

Bronze is an alloy of copper with tin, resulting in a tough, hard metal capable of being hammered or cast into shape. Copper containing 10 percent of tin is initially two-and-one-half times as hard as pure copper; work-hardening by hammering will make the alloy more than six times as hard as the pure metal, a hardness approaching that of forged mild steel (Coghlan 1951 : 44). The rise of this metal gives a name to the next cultural period, the Bronze Age.

Many tools which formerly had been made in copper were now made of bronze. However, there was neither instant nor wholesale conversion to the new metal. Existing, satisfactory items continued to be used. Bronze was expensive and in short supply. The new metal was introduced in the higher, richer segments of society. Bronze was early used for weapons and then only by selected warriors rather than for soldiery in general. The rest of society got along with what they already had for a long time after bronze appeared.

At this point we must recall that the progression of tool types and materials was more a matter of cumulation than of replacement. Some chipped stone tools of the Paleolithic, axes for example, were replaced by ground stone Neolithic models, but many of the chipped types—such as knives and projectile points—continued to be made. The advent of the Copper Age saw the addition of copper rather than the disappearance of stone. And so on. The innovations, as they appeared, were commanded by the rich and the influential. The peasants continued to use tools of long-established types and materials. People of the lower classes and of marginal ethnic groups were evidently aware of their relative deprivation and sought to imitate their betters. Stone imitations, both chipped and ground, of metal artifacts are not uncommon in northwestern Europe. A chipped stone replica of a bronze dagger, dated to the middle of the second millennium B.C., was found in central Denmark, while ground stone axes, which include even the casting marks of the originals, modelled on bronze shaft-hole axes of the early second millennium B.C., were found in Sweden (Powell 1966: 124-125).

The advent of bronze quickened the movements of people in two important ways. Bronze weapons enhanced the ability of their possessors to make war. The movements of conquerors, defenders, refugees, and the conquered created social turbulence and offered opportunities for the interchange of ideas. The quest for bronze, whose initial occurrence is restricted because of the scarceness of tin, led to adventuring, trade, and conquest. In prehistoric Europe the tin mines of Cornwall, the Breton Peninsula, Iberia, and central Europe (Czechoslovakia, Hungary) were the focus of attention (Clark 1952: 194-195).

Again accumulation of goods, garnered mainly through political means, increased. People had permanent residences where goods could be stored, even hoarded. Structures were elaborated and now included substantial numbers of public buildings. The temples, palaces, and tombs of Mesopotamia and Egypt are well known to us; the bulk of these were at least begun in the Bronze Age of each area although additions were made later. Labor was directed toward production of prestige goods for the governing classes. The grave goods of Egyptian Pharaohs are famous. Less well known are the riches of early Celtic chieftains of La Tene culture—torcs and cups of gold, bronze flagons and horse trappings, glass beads, stone sculptures—reflecting the successes of expansionism (Powell 1966: 185 ff.). Life for the privileged acquired a certain barbaric splendor, but for the peasantry it may well have meant an added burden to support. One must remember that the entire structure rested, ultimately, on an agricultural base. The food, fiber, oil, and animal product surpluses were gained from their producers as gifts, tribute, rents, taxes, or the proceeds of extortion. In many respects the details of peasant life were continued without change from the Neolithic.

Anthropologists have come to divide the culture of some times and places into two aspects: a Great Tradition and a Little Tradition (Redfield 1956: 41 ff.; Foster 1967). The Great Tradition is the culture of the elite, the rulers, and consists in such matters as elaborated political organization, organized systems of theology and a contributing priesthood, writing and literature, and fine arts. The accounts of historians, and the objects of much archeological attention, are usually those of the Great Tradition of that culture. The Little Tradition is that of the folk or the peasantry (without using these terms in a technical anthropological sense). These are the common people who live in small houses, till the fields, herd the animals, schedule their planting by the phases of the moon or other augury, and are born, mature, reproduce, and die without lasting notice. In the aggregate they are essential to the culture; individually they are unimportant.

The lines between the two Traditions are not sharp. The servants of the elite were drawn from the folk and gave some indoctrination to elite children. Other nominal folk members were craftsmen who occupied a position midway between the two worlds. The folk themselves carried on an imitation of the ways of the elite, limited to their perception of elite behavior and belief (Wolf 1966).

Finally, the metals of the Bronze Age were widely sought and widely diffused because they conferred on their possessors obvious military and craft advantages. Trade was the principal avenue through which bronze goods were acquired. The sources of ores, especially of tin, were few, and those who held them profited greatly thereby. Although metal items were reworked locally, as they had been since the early Copper Age, the manufacture of bronze artifacts occurred primarily where the ores were smelted and alloyed. The finished goods, not the raw metals, were traded. For example, double-looped bronze palstaves (a type of axhead having loops by which it is fastened to its handle) cast of local tin and copper in northwestern Spain and northern Portugal were believed to have been traded to points in western Europe, from Sweden in the north through the British Isles and western France to southeastern Spain in the south (Clark 1952: 271). Obviously, this distribution process resulted in a substantial uniformity of bronze artifacts over large areas. This regional homogeneity did not necessarily extend to other items which could be locally made of local materials.

THE IRON AGE

The Iron Age was the last of the prehistoric cultural periods we will survey. In some senses, the Age is still with us although there are those who would nominate some other outstanding cultural achievement, such as atomic energy, to give its name to our times.

The Iron Age came on quickly in that metallurgists already had considerable experience and skill and the populace generally had accepted metals. (If the question of popular acceptance of metals seems strange, please consider the following folk beliefs: Iron plows are thought to poison the soil. Some sauces should not be stirred with a metal spoon. Aluminum cooking utensils may poison some foods cooked therein. All of these views, and more, are held by people living in the twentieth century.) Patterns of production and use were established. The lag in development of iron metallurgy is attributable to the general lack of metallic iron in nature and the high temperature needed to smelt and work iron.

Metallic iron in nature is derived from iron meteorites. This metal, a nickel-iron alloy, is a form of stainless steel; had it been simply iron the meteorites would long ago have changed to lumps of useless rust. Although a survey has shown a total of 250 tons of meteoritic iron available, the material occurs only sporadically and unpredictably. The metal was prized when found and usually shaped by cold-hammering. In the absence of the use of heat, this working of the metal must be deemed nonmetallurgical (Tylecote 1962: 9-13).

A critical skill in metallurgy is the ability to generate sustained high temperatures. Almost everything worthwhile which happens to metal occurs either as it is heated or cooled. The common nonferrous metals have comparatively low melting points. Tin has an especially low melting point (232°C), and one can readily understand how it might have come into early use to alloy with copper. Iron, on the other hand, has a melting point (1200 to 1500°C) which is higher than that of copper (1000 to 1080°C); the exact figures depend on the purity and type of metal. Whatever the particulars of the case, early metallurgists were not able to heat furnaces to produce quantities of molten iron. It took the later development of bellows, charcoal, coal, and finally of coke to see full realization of iron metallurgy.

Again, the new metal gave a forward impetus to culture, for it conferred an advantage on its possessors. However, it was not as much of an improvement over bronze as bronze had been over copper. For example, work-hardened bronze (90 percent copper, 10 percent tin) is more than six times as hard as pure cast copper and is close to the hardness of forged mild steel or even medium steel (0.45 percent carbon). High carbon steel (1.25 percent carbon), not an early product, is only one-and-one-half times as hard as that bronze (Coghlan 1951: 44). The Iron Age metal, which was really varieties of steel rather than the pure iron, was somewhat better than existing metals, but it had one major drawback. It rusted. The corrosion products on copper and bronze were not as destructive because they formed a skin over the metal which inhibited further deterioration. Rust just gnawed deeper and deeper into the iron.

Another problem with iron was that it could not be cast with the existing techniques of early metallurgy. (Casting of bronze, especially, was a major metal-forming technique.) The smelters and forges could only produce "point melting," in which little droplets of liquid iron were formed. Cast iron was sporadically produced by Roman metal workers but seems not to have been a regular product. Forbes considers Chinese cast iron to have priority, being well established in the later Han dynasty (25 to 220 A.D.) (Forbes 1950: 407, 440).

Control of value and trade in bronze was achieved through possession of the basic resources at the mines. Without tin supplies one could not get very far in the bronze trade. However, this technique was worthless to control iron trading because iron ores were widely found. The key to monopoly in this instance was found in knowledge—knowledge of the smelting and working techniques for iron. Although there is broad sharing of metallurgical techniques among craftsmen in the various metals, the need for high temperatures seems to set iron a bit apart. Whatever their true understanding of the case may have been, the iron workers let others believe that they were aided by occult powers (Forbes 1950: 62-104). This pose may have discouraged experimentation by the uninitiated and so tended to preserve the mystery of the craft.

The technologies of the Iron Age come surprisingly close to those of recent centuries before the Industrial Revolution. The tools and techniques of the carpenter, mason, blacksmith, cooper, and other craftsmen of the pre-Roman period, for example, can even be identified by traditional artisans of the late nineteenth century.

For all of this apparent modernity we must have a care to distinguish between the Great and Little Traditions. The Iron Age came to the urban dweller, to the cosmopolite, to the people in the center, but virtually passed by the folk without notice. People in the backwoods, on the peripheries, in the lower classes, continued to live in a style which could be termed "Neolithic with metals added." Metals were not so thoroughly integrated into their existence as to produce an unequivocal dependence. Cutting tools were unquestionably their most important use of metal and the one least likely to be replaced with some other material (such as with stone). Ceramic, stone, wooden, and basketry containers had hardly given way to metal containers. Wooden structures were fastened by fitting pieces closely, by pegging, and by lashing rather than with metal. One may cite still other cultural practices in which metal, still an essentially scarce material, was not habitually used by folk peoples.

A peasantry is sometimes defined as the rural aspect of a complex society. The peasants often have a culture which is self-contained in all save a few important aspects. They may have local building and dress styles, local food specialities, local festivals, local deities, and a localized kinship structure. Their villages are, with respect to internal affairs, virtually autonomous under the leadership of a chief or a council. However, the external relations of villages and a large part of their economies are managed by an urban governing elite who may live at a distance. The peasants may be called upon to support or fight in wars about which they know little. In return they can expect to be defended against foreign attack or, minimally, led in self-defense efforts. In the economic sphere the peasants found their production was dictated from outside the community. They were told how much of what crop to deliver as land rent or taxes. It might turn out that they were forced to devote all their land and time to a cash crop so that their food was, perforce, purchased from others (Wolf 1966; Redfield 1953). In medieval times the peasant had to have his grain ground at the landlord's mill (perhaps as a check on his crop) so that home milling equipment, such as the quern or mortar and pestle, became uncommon (Hall 1962; Handlin 1967: 462). The controlled economy is not a modern phenomenon.

In viewing the long haul from the earliest technological activities of man to modern times we must retain our perspective. Though this prehistoric time span is divided into Ages, such division is for the convenience of prehistorians and others. Each Age is described in generalized terms, and these offer an ideal picture of the culture of the times. No man of such a period could ever have suddenly come to the realization that he lived, for example, in the Bronze Age with the same kind of wonder that was encountered by the man who discovered that he spoke in prose. Times passed and cultures changed, with few decisive events which, at a stroke, revolutionized the life of a place or a time.

The termination of this account must be on some arbitrary basis for there was no stopping of cultural development at any point. The author, reflecting the bias of some anthropologists, would be inclined to draw the line at the beginning of the historic period. However, this choice means that the cutoff point will vary greatly from one part of the world to another. This historic period began, in a way, several thousand years before Christ in the Near East, but only a century ago in some parts of Africa and South America. The arbitrariness of choice becomes obvious. Also the distinctions between some technological circumstances of slightly pre-Roman times and those of late medieval Europe are small, so that one may use examples rather freely over a range of centuries. Consequently, we will stop with the advent of history, but keep our options open nonetheless.

The Dynamics of Material Culture

Material culture and technology, like all the rest of culture, are changing. Because they are a part of culture, their dynamics may be examined in the same manner as the rest of culture. The most significant difference between the consideration of this pair and of other cultural manifestations is that material culture and technology, almost by definition, have left us with a very long record. Unfortunately, this long record is not as useful to us as might be supposed because it does not reveal many of the imponderables of human behavior such as attitudes and motivations. Let us first direct our attention to other aspects of technological dynamics and postpone some of the more difficult questions.

Each technological circumstance depends upon prerequisites in knowledge and in skill. Fine products reflect, even to the uninitiated eye, the skill of their maker. And few people would expect their products to be of comparable quality unless they, too, had had opportunities for similar instruction and experience in the craft. However, what is not so evident is that much simpler and cruder works also depend upon knowledge and skill. Even the knowledge that stones may be chipped into desired shapes is not innate to any man. It has to be discovered, probably many times, and put into practice. The skill to chip stones into tools, at any level of refinement, is acquired through experience and a lot of bruised fingers and scraped knuckles. If the reader is in any way convinced that stone chipping is easy, let him try. A little bit of information, a minor skill underlies everything man makes.

Not only are there prerequisites to the manufacture of individual products, but there are antecedents to whole cultural stages. A distinction was drawn, following anthropological practice, between horticulture and agriculture. The horticultural development called for the presence of domesticated plants cultivable with hand tools. Agriculture called for domesticated plants of the same or a different kind plus the domesticated animals which could pull the plow. Although the evidence is not conclusive, it is probable that domestic, or cultivated, plants were used by man before domestic animals (Protsch and Berger 1973). So, logically, horticulture would precede agriculture simply because it requires one less element.

This argument of building one block upon another is in large measure responsible for the attractiveness of cultural evolutionary theory. In retrospect at least, one may see that one event led to another. Because of the prerequisite aspect this approach has a certain neatness about it.

Two closely related questions are left with partial answers. First, given the existence of cultural traits from which another step might be made, how long will it be before that step is actually taken? For example, given the appropriate plants and given plow-capable animals, how long is it until the two are conjoined in the practice of agriculture? Second, is such a progression inevitable? Hindsight "prediction" is easy, but cultural forecasting is most difficult. To engage in prediction of this kind involves actually being the innovator, to be capable of recognizing fruitful combinations which may be made of existing traits. That is a large order.

Some consideration of a closely related issue has been made by two social scientists (Ogburn 1938: 90-102; Kroeber 1948: 341-342, 364-367). Their problem revolved around the occurrence of simultaneous inventions, those which were brought forward at substantially the same time by individuals who were not known to have been in contact. In a substantial number of cases, the inventions came out of the same, or markedly similar, cultural contexts. To this extent the inventors were in communication. The conclusion to which one is led by their investigations is that there is a measure of inevitability about inventions, although we remain uncertain about the time span of element availability preceding the creative acts.

CHANGE AND CONTINUITY

At several points in our discussion stress has been laid simultaneously on change and continuity. These two concepts must be considered as two manifestations of the same thing rather than as two mutually exclusive ideas. Material culture reflects the oneness of these twin concepts better than most aspects of culture.

A new machine is built. Its form and function need not concern us here. (Just imagine a machine, any machine.) The frame is of cast iron. Cast iron has been around for more than a thousand years. Its gears are mostly steel, even older than cast iron, but the other parts are made of a new synthetic material—tough, self-lubricating, long-wearing—which quiet a formerly noisy device. The mechanical motions involved in the machine are conventional and may be found in any engineer's handbook. However, the controls depend upon magnetically responsive fluids which operate clutches. And so on. All "new" items involve the old as well as the new.

Change results from forces both internal and external to the individual and to his culture. Members of a culture may suffer from boredom and seek new ways to excite a jaded psyche. Boredom and the resulting quest for new stimuli are evident in the recreational scene in the contemporary United States. Where are yesterday's go-carts or slot-cars? What happened to canasta? A similar ennui seems to overcome conventional political life from time to time and results in splinter parties, in reform candidates, and in populist demagogues. While one may consider changes of this kind as mere fads, soon to be forgotten, this view is erroneous. Each of these innovations leaves its mark on the culture even though it is ostensibly forgotten and others remain to grow influential. Remember that aviation began as a scientific curiosity and gained its first momentum as a rich man's sport.

Changes stemming from external forces include the results of conquest, more peaceful contacts with other people, and environmental shifts. One of the common consequences of conquest is the imposition of a new language, that of the conquerors. The economy is often altered in ways similar to those mentioned above in the discussion of a peasantry. Economic activities are reflected in material goods—the tools associated with new crops, the disappearance or relegation to secondary status of items belonging to an abandoned hunting-gathering subsistence, the diminution in prestige goods as surpluses are siphoned off as tribute. The conquerors may impose rules governing the dress and deportment of their subjects, rules which prohibit fancy dress or elaborate houses or the use of horses or carriages.

Peaceful contacts have been those of trade and intermarriage. Trade is responsible for the occurrence of "exotic" goods, as the archeologist sometimes calls them. These items, not of local manufacture, are often identifiable as imports because they stem from technologies not practiced in the area (meaning that the products, but not the means to make them, are found archeologically) or because they reflect raw materials which are also alien to the area. Intermarriage offers a special source of cultural change. When intergroup marriages occur they are usually not fully reciprocal in that one sex or the other is mobile and not both equally. Consequently one notes the movement of techniques and their products, not the products alone.

Crafts are always sex-associated (for example, men do most stoneworking), especially in preindustrial societies. With the movement of men, or of women, in marriage from one society to another there would be the corresponding transfer of crafts practiced by members of that sex. This transfer would be evident in that one sex's kind of material culture would be closely shared between the interchanging societies while other material things would be societally different. For example, if two groups intermarried with the women going to live in their husband's tribe, it might be found that the pottery (made by women) was relatively homogeneous while woodenware (made by men) differed between the groups.

Lastly, environmental changes come to people in two ways: Those which occur as they stay in one location and those which result from migration. Both circumstances produce technological changes. The ending of the Pleistocene and the retreat of its glaciers wrought a powerful change in the life of Mesolithic man. Some groups migrated in an effort to stay with the climatic zone to which they were accustomed and adapted. Other groups developed new adaptations, new cultural orientations, to fit their altered surroundings. A recent parallel to these circumstances and responses may be found in the events of the 1930s when the Dust Bowl of the Middle West forced change on the inhabitants of that area. Many abandoned farming, some to migrate westward, others to try their hands at new occupations in town. Even for those who remained there were alterations in the mode of farming to cope with drought and prevent a recurrence of airborne soil loss.

Anthropologists have focussed their attention on cultural change to the relative neglect of cultural continuity. Continuity is not simply the lack of change. There are positive forces for continuity just as there are for change.

The entire maturation process, including both informal and formal education, encourages continuity in culture. The learner learns that which is presented to him. The teacher teaches that which he knows. The master teaches his apprentice. The same occurs between parent and child. Anthropologists call the total process "enculturation," in other words, the induction of a new member of a group (the child) into the ways of the group's culture. It is a culture-conserving mechanism.

Isolation, either geographic or attitudinal, favors cultural continuity. If a group is in no position, literally or figuratively, to receive new ideas from outside, then their culture persists with only minor changes. The peoples of Tierra del Fuego, at the southern tip of South America, are known as having been quite primitive. Their location put them far from the mainstream of new ideas well to the north. Admittedly their environment was a tough one, but they surely could have profited to some degree from more outside contacts. The Tasmanian natives, whose environment offers no such stringent limitation, offer a parallel instance of an extremely isolated people of considerably primitive culture. Sometimes a society will have such a high opinion of its own worth as to find it impossible to accept ideas from any other group because all foreigners are ipso facto inferior. It has been said that attitudinal isolation was true of China around the time of Marco Polo.

Habituation favors continuity. There is the habituation of familiarity in that the known is comfortable, predictable, and reassuring. We live in a world of minor expectations. Sometimes we understand this best when an expectation is not realized. The study in which I write this page has a ceiling light controlled by a wall switch. When you walk into the darkened room and reach for the switch it seems to have vanished. You grope around. Perhaps there is no wall switch. Ah, here it is! The switch is thirty-eight inches from the floor instead of the customary forty-six. Too low, you may think. However, the switch was really at the "right" height if you know that the floor was raised nine inches after the switch was installed. What set the height of wall switches in the first place?

The illustration above pertains to those culture-stabilizing behaviors called motor habit patterns. These patterns are the organization of movement ("motor" here refers to motion, movement). We become habituated to making certain kinds of physical movements in the manipulation of things, such as the level at which you reach out for a switch. One of the most habituated of these movements is writing. Consider, if you will, the stability of handwriting in a signature or even in common handwriting of an individual through the years. Suppose we try to alter our handwriting. Some of you can doubtless "mirror write," from right to left as a complete mirror image of normal writing. Have you tried writing upside down from left to right or even upside down mirror writing? Remember that you can read fairly well upside down. Why, then, not write with equal facility that way? A part of the answer to this question doubtless lies in the differences between the motor habits of writing and those of reading. In the latter, the eye moves along to view whole words or phrases, not to laboriously trace each line of each letter. The driving of an automobile becomes so habitual that the experienced driver gives virtually no conscious thought to the physical aspect of controlling the vehicle, his efforts are largely judgmental with reference to traffic, obstacles, his route, and related matters.

Finally, change does not alter the entirety of culture even in revolutionary times. In general, the bulk of a culture's content remains unchanged over long spans of time. In this way continuity overshadows change.

INNOVATION AND DIFFUSION

New cultural things arise from the two closely related processes of innovation and of diffusion (or borrowing).

We popularly distinguish between invention and discovery, but the grounds for the distinction are not always demonstrable. An invention may be defined as something created anew which did not exist before the act of invention, the product of deliberation, of a purposive search. An invention may be the new application of existing knowledge. By contrast with each of the foregoing, a discovery is the revelation of something which had prior existence but was unknown. A discovery is a chance addition to the stock of knowledge. Only with the most recent events, in which we can fully investigate the situation and interview the inventors/discoverers, can we make some of the decisions needed to distinguish invention from discovery. What evidence have we, from the remote past, of the existing stock of knowledge? What can we know of the inventor's motivation? The most important fact about an invention or discovery is that it is new. Consequently, the word "innovation" has been proposed to take the place of both (Barnett 1953: 7-10).

In discussions of the causes of innovation anthropologists have traditionally spoken in terms of "needs." According to the conventional wisdom, these needs—whether biological, social, or psychic—have given rise to responses in the form of innovations. Some of this reasoning must be granted as true. However, when one considers how many peoples of the world live, and have lived, in wretched conditions of which they are fully aware, it is difficult to see anything spontaneous in the need-response relationship. It seems that the ultimate sources of innovation have not been revealed.

Many students of culture take a deadly serious view of their subject. They think and write soberly of human needs and their fulfillment. For them, as for Longfellow, "Life is real! Life is earnest." With this view they tend to overlook the lighter side of life. Apparently play is a potent source of innovation. A desire for novelty is well demonstrated. A desire for relief from routine may lead to new ideas. Gearing to automatically open and close valves on a steam engine (first seen on a Newcomen engine) may be traced to one Humphrey Potter who sought to relieve the drudgery of controlling the valves manually as had formerly been the case (Mantoux 1962: 317).

Skilled craftsmen play with their craft—trying new approaches, endeavoring to surpass customary standards of work quality, demonstrating their virtuosity. Out of this self-entertainment come new ideas which spread to other craftsmen and to other cultures. Play as a motivation for human activity is not to be set aside as mere frivolity. Recent studies of nonhuman primates suggest that a considerable part of their time, at least in captivity, is taken up with play which involves the manipulation of things (Poirier 1970: 339-341). Isn't it possible that this behavior was an element in the first primate uses of tools?

The history of cultures has been accompanied by an increase in cultural change. Innovation has been one source of change. The number of innovations is seen to increase. Although the absolute rate (the number of innovations per person) may not have altered, there are more people than ever before, hence more innovations. This impression may be gained because recent events always loom larger than more remote ones, but there is strong reason to believe that no aberration exists. To what may we attribute this increase?

All new cultural things are made, in part, of old cultural things. It may be that an innovation lies solely in the nature of the combination, that the integration is novel. (It is sometimes of such ideas that we ask ourselves, "Now why didn't I think of that?" For example, the pencil eraser which is alleged to have earned its innovator a fortune.) Because innovations incorporate some prior cultural elements, it stands to reason that the more elements a culture contains the more innovative combinations potentially exist. The acceleration of innovation in recent times is deemed to be a product of having more elements available to recombine. This is another instance in which "the rich get richer."

A twin to innovation is diffusion, the spread of cultural elements from one culture to another. In effect, one culture lives in part off the other's innovations. The extent to which diffusion is important in the building of cultures was the topic of heated anthropological debate with the cultural evolutionists (essentially innovationists) ranged against the diffusionists (Dixon 1928; Lowie 1937). A truce, but not a final settlement, has quieted the argument for some decades now.

Diffusion operating interculturally has much the same dynamics as the spread of innovations within a culture. An important exception to this parallel is that rejection of an innovation means that it is lost, for the time being, to all cultures (Rogers and Shoemaker 1971: 8-12).

Diffusion may seem a simple process. The people of a culture possess a trait—an artifact, a belief, a behavior—about which the people of another culture learn by contact, trade, or hearsay. As a consequence of this contact, the trait is adopted in a new context. However, there are many complicating factors. In the first place, the potential recipients must be attracted by the trait; they must see it as being of value to them. Importantly, they accept the trait on their own terms, not those of the donor culture. Let us illustrate this point with a hypothetical, but not far-fetched, example. Group A is industrious and hardworking. They believe that it is meritorious to rise early and complete their work in the cool of the day. They have alarm clocks to wake them at an early hour. Group B is also industrious and hardworking, but they reject the alarm clocks offered by Group A. Group B believes that one should not abruptly wake a sleeper lest his soul, perhaps wandering in a dream, not have time to return. The alarm clock, in the view of Group B, is not an aid to diligence but a threat to the integrity of one's soul.

An item which is diffused may be known to its receptors only for its physical nature, its connotations in the donor culture having been lost in transmission. A new set of values and behaviors may be built up around the new trait. For example, when World War II brought the hand grenade to the southwest Pacific islands, it was quickly used by the natives to stun fish in lagoons and rivers. The behavior and values of this usage were quite different from those of use in warfare, an activity not unknown to the natives but in which they chose not to employ the grenade. Alternatively, the new trait may be assimilated into an existing behavioral set for a trait deemed analogous. If, for example, passenger trains have been operated by the state instead of private companies, when buses are introduced, they may well also be state operated. Several railway-bus-ferry systems in European countries illustrate this development.

In the early nineteenth century, the Zulu nation under a great leader, Shaka, was engaged in prolonged wars with native neighbors, with the Dutch, and with the British. Firearms, secured through trade and capture, were used in addition to native weapons. Although the Zulus had centuries of experience in smelting and working iron, they were unable to make guns and quite limited in their ability to repair them. Consequently, the Zulus tolerated in their midst some European Christian missionaries who exchanged their services as armorers for an opportunity to proselytize (Oliver and Fage 1962: 163).

Thus it may be seen that cultural items diffused through trade were often received without detailed knowledge of their manufacture. This was often the case when advanced and primitive cultures came together, for the latter lacked the technological sophistication to analyze and imitate the goods they received. However, when technological equals meet it is only a matter of time until the receptor is making the item on his own, assuming that he has the resources. It is probable that his techniques will closely parallel those of the donor culture because, as equals, they share a great many technological capabilities and substantial technological knowledge. The rapid development of subatomic particle accelerators, such as the cyclotron, by many countries in the face of attempts at secrecy illustrates the point. Much the same outcome is apparently taking place in the nuclear explosive and nuclear power field. The delay in discovering the true nature of porcelain, known in Europe for centuries as an import from China, can be explained as due to the lag of European ceramic technology behind that of the Chinese. Once some measure of equality was achieved, then analysis and duplication of the substance were forthcoming (Clow and Clow 1958: 336-341).

Diffusion by hearsay, "stimulus" diffusion, involves the transmission of the essence of an idea. Contact between donor and receptor cultures will be tenuous. The receptor culture, on hearing of the trait, works out the details to suit itself. This freedom, or lack of information, can lead the trait in new directions. Suppose that Group A heard that Group B had a writing machine and decided to adopt the trait. However, Group A lacks printing from movable type so they make their writing machine in a form which moves a pen cursively over the paper instead of a form similar to our typewriter. The minimal knowledge was a stimulus to action, the results are noticeably different.

In summary, diffusion involves contact, appreciation, acceptance, and integration. At each step along the way, people have to make a decision. Although we often speak or write as though the diffusion occurred between cultures, in fact it is between people. Their characteristics, their likes and dislikes, their vision or narrowness all bear on the outcome of the venture.

POSSESSION AND PRODUCTION

The growth of cultures has demonstrated the changing integration of given cultural items. This potential shift exists whether the new trait originated by innovation or by diffusion. The point is readily illustrated by the development which we have already examined in the use of metals. When a metal is first introduced it appears in the hands of a limited number of people, usually an elite, in the form of weapons. This usage has been termed "restricted." As the metal becomes more plentiful it becomes available to a larger segment of the population, ultimately to all of it. Further, its uses diversify so that it is no longer exclusively employed for highly valued artifacts. It has now become, in the same terminological system, a "universal." A third circumstance of use and possession is possible. Let us assume for the moment that the metal remains solely the material of weapons and that the society concludes that it does not want promiscuous possession of weapons by its members. Then the material is said to be the "speciality" of some select segment of the group, perhaps of the police or the military (Linton 1936: 272-274).

Although this simple analytic scheme is not always applicable in every detail, we must grant that it does approximate the facts. In dealing with material culture we find that all items are not necessarily available to everyone. This circumstance is as true of the simpler cultures as it is of the more complex ones.

Production, possession, and ownership of artifacts are universally restricted by custom, the rules of the human game. Among technologically simpler peoples the production of any given class of goods is customarily in the hands of individuals of one sex. This restriction is called a simple or natural division of labor. For example, if weaving is customarily done by women, it would be inappropriate for a man to engage in the craft. However, sometimes this rule is reversed or waived when the product will have ceremonial or religious use because women, often barred from esoteric matters, would be potentially defiling to the product.

In more elaborated circumstances the production of a given item is exclusively in the hands of specialized craftsmen who may be organized into a guild. Their right to produce is protected by societal decree or custom and may be maintained because only they are considered to have the proper relationship to the supernatural world. For an outsider to attempt the same craft would be impious and a courting of disaster for the society as well as the individual. A kosher butcher of today, preparing meats for a Jewish clientele, can be cited to illustrate the principle. His tasks, then, are not something at which any person may try his hand.

In addition to restrictions placed primarily on the producer or possessor of an artifact, there are limits placed on the artifacts themselves as classes of objects. Possession of religious objects is often limited to priests, old men (as among Australian aborigines), or males who have undergone a ceremony functionally similar to confirmation (such as the "twice-born" of Hinduism). Today we except from this rule those religious items which have passed over into the category of objects of art. Yet there is some resistance to this reclassification for it accelerates the stripping of churches to supply the art market.

Private ownership of venerated political symbols and documents is also restricted. For this reason it is quite improbable that the Declaration of Independence or the Liberty Bell would come to be the property of an American collector. The public sense of impropriety, perhaps assuaged by the nationality of the holder, would be greatly heightened if, in time, these national treasures were sold to a Greek shipping magnate. Even the bodies of distinguished dead become "public property," witness French distress at the theft in 1973 of General Henri Philippe Pétain's body from the grave in which it had lain for twenty years.

Some weapons and destructive devices are limited in possession to police and military forces. Even some kinds of tools are considered by the police to be prima facie evidence of unlawful intent. That these restrictions are not paralleled in every detail among primitive and preindustrial peoples reflects only their simpler artifactual inventory and less complex division of labor, not the absence of the general principle.

It is in the nature of culture to be dynamic. Consequently, material culture and technology as special facets of culture are likewise dynamic, a fact often ignored because it is convenient to analyze a static situation. This error of convenience is often coupled with the aberration resulting from a foreshortening of historic perspective. As an instance of the latter, beyond the modern, industrial scene for which we assume (often falsely) a certain sameness, the cultures of the distant past form an apparently homogeneous group. These two shortsightednesses combine to persuade us that recent primitives are unchanged descendants of the past. Certain continuities would lead us to think so. However, any archeologist can tell you that he relies heavily on continuous minor changes in material culture to differentiate cultural levels or horizons in the places he excavates. Stylistic changes in fishnet sinkers, for example, are paralleled by those in pottery, by those in arrow points, and by still others.

Statements that cultures have not changed, and we all make such claims, stem from failure to examine the situation with sufficient care. We may rest assured that cultures change if only because the homeostatic processes which might resist change fail to function perfectly.

Technology in the Modern World

The keynote of the modern world is the complexity and ubiquity of its technology. Hardly any activity among westernized people is conducted with elemental simplicity. The sale of a button is monitored by a computer, a guitar is electrically amplified or even replaced by a wholly electronic device, a stage play may be combined with photo-projection, and religious services are augmented with sound-and-light shows. "Modern" has come to mean "technically complex," often in the style of Rube Goldberg, rather than indicating an elegant simplicity resulting from complete mastery of an art.

Our western world has come to pride itself on the complex state of its technological order. Ethnocentrism gives many westerners a bias in favor of technology and its products. The "underdeveloped" countries are those thought deficient in material goods and the means to make them. Standards of living are measured, in large part, in physical things rather than the less tangible elements of individual enjoyment, freedom of expression, and satisfaction with one's social situation. (Lately, one must admit, we have come to recognize another measure of human existence called "the quality of life," which apparently takes for granted the substantial fulfillment of man's physical needs.) Achievements in other aspects of culture—for example, in domestic relations, in political or legal structures, in literary expression—are deemed of lesser value than technological ones. Instead one hears praises sung for artificial turf, the push-button telephone, four-channel sound, and the automobile "bumper" which will withstand a five-mile-per-hour crash.

The bias toward technology is well documented in a recent and insightful article on the manner in which technology is being brought to bear on social problems (Etzioni and Remp 1972). Two sociologists discuss the methadone program for heroin addicts, the employment of antabuse in fighting alcoholism, instructional television as an alternative to "live" teaching, the breath analyzer in handling drunken drivers, and similar cases. Note that technological means are being employed in the modification of behavior when we have more customarily made use of social or psychological influences for this purpose. Moreover, one may see that the technological approach is totally external to the individual and may be applied, perhaps against his will, by outside agencies.

When the bias is revealed and the technocentric view is challenged, the adherents will base their defense not on the central premise but on its peripheral effects. For example, military research and development have not only increased the scope and degree of martial devastation, but have contributed to perfection of computers, radiation therapy, heat-resistant materials, and electronic miniaturization. While these products may be desirable, one may ask if there is not another way in which to achieve them. The analogous question might be asked, whether it was necessary to have a Fascist government in Italy in the 1930s in order to have the trains run on time?

Technology has a tendency to feed upon itself and so increase in content and complexity. Undoubtedly a part of the tendency is attributable to the same force that motivates mountain climbers: "Because it's there." Technicians generally have chosen their profession because they enjoy working with technical things. They have learned to think in a technical vein and to seek technical solutions rather than solutions of other kinds. (Wouldn't it be a bit out of character for a consulting civil engineer to suggest to a client that prayer would be the best response to the problem of a bridge grown dangerous by advancing age?) There is an element of pride and an element of play in what they do, from which they must gain some personal satisfaction. Can one blame them?

There is a notable predilection to pile one device upon another in quest for solutions to problems rather than rethink the entire situation. Attempts to meet increasingly stringent federal standards for pollution control in automobiles are replete with tinkering in all aspects of the machine from the air intake to the exhaust pipe and parallel experiments with gasoline specifications. For a variety of reasons, only some of which are economic, consideration of radical departures from the conventional internal combustion engine has been delayed until recently. At that, many alternatives now being mentioned have been operated, often quite successfully, at times during the past fifty years! The steam engine, the turbine engine, and the Stirling-cycle engine are old standards even though they lacked the popularity of the Otto-cycle engine now widely used.

The additive approach is also observable in one of the technological "solutions" to a social (medical?) problem mentioned above. Methadone, used for treatment of heroin addiction, is itself addictive and subject to abuse. Lethal overdoses can be generated from the amounts furnished to patients. The addition of another drug, naloxone, in small quantities (down to 1 percent) is now suggested to stop overdosage of methadone and the black market traffic in this substance. Naloxone may also prevent misuse of paregoric and make possible a resumption of over-the-counter sale of that drug (Pachter 1973).

The technological approach has been criticized on at least two major grounds. In the first instance, many technological operations produce noxious by-products which foul the environment. The automobile is evidently the major producer of smog. Flue gases from smelters have denuded the adjacent countryside. The effluent from paper mills has "killed" streams. A second criticism has been the high cost in natural resources exacted by elaborated technologies. Most of the resources we use are naturally regenerated, but at rates far removed from present rates of consumption. Perhaps the resource most closely paced, or potentially renewable, with our consumption is the wood from forests. Trees do grow to usable dimension in less than a human lifetime, which is more than may be said of coal or petroleum deposits or of copper ore. As we promote technologies which have high costs in resource use, we are spending the inheritances of our grandchildren and their descendants. Hardly any forecaster offers other than a pessimistic view of the future for basic resources; the differences in timetables for the exhaustion of any given mineral or fuel are only on the order of a few decades—not enough to offer any long-range hope.

Distress over this technological emphasis and over related environmental abuse has led in recent years to the rise of counter movements. The extrication of this aspect of protest from other ideological concerns is not easily accomplished, yet may be feasible in some cases. The protesters have marched under so many banners that no one label suffices to identify them. Among the elements of their activities one may find communal rural living, organic gardening, recreational backpacking, and the resurgence of bicycling. (Doesn't this sound like an "underdeveloped" country?) Virtually all of these people wish to tread as lightly as possible on their environment, to minimize their impact. With all their good intentions a reliance still is placed on manufactured goods, especially those of an automotive and electronic nature.

The foregoing criticism should not be interpreted as support for unbridled technological operations. The author feels that the world is too valuable, too unique, to be left to the care of ungoverned technologists. A balance must be struck between the needs of society, the capabilities of technologies, and the resources which are to supply both. It has been emphasized that the answer to technologically generated problems is not more technology. A counter-technological viewpoint must exist to promote other kinds of solutions to these problems and, perhaps, to reduce their frequency and severity in the first instance. If we do not take this position then we will find ourselves in the situation of the cigaret smoker who is confident that a cure for cancer will be found before he develops the disease.

Literature Cited

H.G. Barnett. *Innovation, the Basis of Cultural Change* (New York, 1953).

Lewis R. Binford. "Post-Pleistocene Adaptations." In Stuart Struever, ed. *Prehistoric Agriculture* (Garden City, 1971) pp. 22-49.

Jacques Bordaz. *Tools of the Old and New Stone Age* (Garden City, 1970).

Francois Bordes. *The Old Stone Age* (New York, 1968).

Robert J. Braidwood. *Prehistoric Men* (7th ed.; Glenview, 1967).

V. Gordon Childe. *The Dawn of European Civilization* (New York, 1925).

Graham Clark. *Archaeology and Society* (Reprint of 3rd ed., 1957; New York, 1964).

J.G.D. Clark. *Prehistoric Europe: The Economic Basis* (New York, 1952).

A. and N.L. Clow. "Ceramics from the Fifteenth Century to the Rise of the Staffordshire Potteries." In Charles Singer et al., eds. *A History of Technology* (New York, 1958) vol. 4, pp. 328-357.

H.H. Coghlan. *Notes on the Prehistoric Metallurgy of Copper and Bronze in the Old World* (Oxford, 1951).

Don E. Crabtree. "Flaking Stone with Wooden Implements." *Science* 169 (1970) 146-153.

E. Cecil Curwen and Gudmund Hatt. *Plough and Pasture* (New York, 1953).

Raymond Dart. *Adventures with the Missing Link* (New York, 1959).

Roland B. Dixon. *The Building of Cultures* (New York, 1928).

Philip Drucker. *Cultures of the North Pacific Coast* (San Francisco, 1965).

Amitai Etzioni and Richard Remp. "Technological 'Shortcuts' to Social Change." *Science* 175 (1972) 31-38.

R.J. Forbes. *Metallurgy in Antiquity* (Leiden, 1950).

C. Daryll Forde. *Habitat, Economy and Society* (5th ed.; New York, 1963).

George M. Foster. "Introduction: What is a Peasant?" In Jack M. Potter et al. *Peasant Society* (Boston, 1967) pp. 2-14.

A. Rupert Hall. "The Changing Technical Act." *Technology and Culture* 3 (1962) 501-515.

Oscar Handlin. "Peasant Origins." In George Dalton, ed. *Tribal and Peasant Economies* (Garden City, 1967) pp. 456-478.

Jesse Jennings. *Prehistory of North America* (New York, 1968).

Charles M. Keller. "The Development of Edge Damage Patterns on Stone Tools." *Man* 1 (1966) 501-511.

Fritz L. Kramer. *Breaking Ground: Notes on the Distribution of Some Simple Tillage Tools* (Sacramento, 1966).

A.L. Kroeber. *Anthropology* (2nd ed.; New York, 1948).

Herbert Kühn. *On the Track of Prehistoric Man* (New York, 1955).

Annette Laming. *Lascaux* (Baltimore, 1959).

L.S.B. Leakey. "Working Stone, Bone, and Wood." In Charles Singer et al., eds. *A History of Technology* (New York, 1954) vol. 1, pp. 128-143.

Ralph Linton. *The Study of Man* (New York, 1936).

Robert H. Lowie. *The History of Ethnological Theory* (New York, 1937).

Paul Mantoux. *The Industrial Revolution in the Eighteenth Century* (New York, 1962).

Paul Martin, George I. Quimby, and Donald Collier. *Indians before Columbus* (Chicago, 1947).

Kenneth P. Oakley. *Man the Tool-Maker* (6th ed.; London, 1972).

William F. Ogburn. *Social Change* (New York, 1938).

Roland Oliver and J.D. Fage. *A Short History of Africa* (Baltimore, 1962).

Irwin J. Pachter. Letter. *Science* 179 (1973) 230.

John E. Pfeiffer. *The Emergence of Man* (New York, 1969).

Frank E. Poirier. "The Nilgiri Langur *(Presbytis johnii)* of South India." In Leonard A. Rosenblum, ed. *Primate Behavior* (New York, 1970) vol. 1, pp. 251-383.

T.G.E. Powell. *Prehistoric Art* (New York, 1966).

Reiner Protsch and Rainer Berger. "Earliest Radiocarbon Dates for Domesticated Animals." *Science* 179 (1973) 235-239.

Barbara Purdy and H.K. Brooks. "Thermal Alteration of Silica Materials: An Archeological Approach." *Science* 173 (1971) 322-325.

Robert Redfield. *Peasant Society and Culture* (Chicago, 1956).

_____. *The Primitive World and its Transformations* (Ithaca, 1953).

Everett M. Rogers and M. Floyd Shoemaker. *Communication of Innovations* (2nd ed.; New York, 1971).

Marshall D. Sahlins and Elman R. Service, eds. *Evolution and Culture* (Ann Arbor, 1960).

S.A. Semenov. *Prehistoric Technology* (London, 1964; original Russian, 1957).

R.F. Tylecote. *Metallurgy in Archaeology* (London, 1962).
Theodore A. Wertime. "Man's First Encounters with Metallurgy." *Science* 146 (1964) 1257-1267.
Leslie A. White. *The Science of Culture* (New York, 1949).
Eric R. Wolf. *Peasants* (Englewood Cliffs, 1966).